Rastignac the devil

by

Philip Jose Farmer

Double9 BOOKS

Rastignac the devil

by Philip Jose Farmer

ISBN: 978-93-59328-13-3

Published by

DOUBLE 9 BOOKS

2/13-B, Ansari Road
Daryaganj, New Delhi – 110002
info@double9books.com
www.double9books.com
Tel. 011-40042856

ABOUT THE AUTHOR

Philip José Farmer (January 26, 1918 – February 25, 2009) was a science fiction and fantasy novelist and short story writer from the United States. Farmer is most known for his novel series, particularly his World of Tiers (1965-93) and Riverworld (1971-83) series. He is known for his groundbreaking use of sexual and religious themes in his work, as well as his obsession with and reworking of the legend of celebrated pulp heroes, and for his occasional tongue-in-cheek pseudonymous works written as though by imaginary people. Farmer frequently blended actual and classic fictional characters and settings, as seen by his Wold Newton family stories, which connect classic fictional characters together as real people and blood relatives as a result of an alien conspiracy. Early examples of literary mashup novels are The Other Log of Phileas Fogg (1973) and Doc Savage: His Apocalyptic Life (1973). Leslie Fiedler, a literary critic, compared Farmer to Ray Bradbury, describing both as "provincial American eccentrics" who "strain at the classic limits of the [science fiction] form," but Farmer stood out due to his capacity to "be at once naive and sophisticated with his odd blending of theology, inappropriate material, and adventure."

CONTENTS

Enslaved by a triangular powered despotism—one lone man sets his sights to the Six Bright Stars and eventual freedom of his world.

After the Apocalyptic War, the decimated remnants of the French huddled in the Loire Valley were gradually squeezed between two new and growing nations. The Colossus to the north was unfriendly and obviously intended to absorb the little New France. The Colossus to the south was friendly and offered to take the weak state into its confederation of republics as a full partner.

A number of proud and independent French citizens feared that even the latter alternative meant the eventual transmutation of their tongue, religion and nationality into those of their southern neighbor. Seeking a way of salvation, they built six huge space-ships that would hold thirty thousand people, most of whom would be in deep freeze until they reached their destination. The six vessels then set off into interstellar space to find a planet that would be as much like Earth as possible.

That was in the 22nd Century. Over three hundred and fifty years passed before Earth heard of them again. However, we are not here concerned with the home world but with the story of a man of that pioneer group who wanted to leave the New Gaul and sail again to the stars....

Rastignac had no Skin. He was, nevertheless, happier than he had been since the age of five.

He was as happy as a man can be who lives deep under the ground. Underground organizations are often under the ground. They are formed into cells. Cell Number One usually contains the leader of the underground.

Jean-Jacques Rastignac, chief of the Legal Underground of the Kingdom of L'Bawpfey, was literally in a cell beneath the surface of the earth. He was in jail.

For a dungeon, it wasn't bad. He had two cells. One was deep inside the building proper, built into the wall so that he could sit in it when he wanted to retreat from the sun or the rain. The adjoining cell was at the bottom of a well whose top was covered with a grille of thin steel bars. Here he spent most of his waking hours. Forced to look upwards if he wanted to see the sky or the stars, Rastignac suffered from a chronic stiff neck.

Several times during the day he had visitors. They were allowed to bend over the grille and talk down to him. A guard, one of the King's mucketeers, [1] stood by as a censor.

[1] Mucketeer is the best translation of the 26th century French noun *foutriquet*, pronounced *vfeutwikey*.

When night came, Rastignac ate the meal let down by ropes on a platform. Then another of the King's mucketeers stood by with drawn épée until he had finished eating. When the tray was pulled back up and the grille lowered and locked, the mucketeer marched off with the turnkey.

Rastignac sharpened his wit by calling a few choice insults to the night guard, then went into the cell inside the wall and lay down to take a nap. Later, he would rise and pace back and forth like a caged tiger. Now and then he would stop and look upwards, scan the stars, hunch his shoulders and resume his savage circuit of the cell. But the time would come when he would stand statue-still. Nothing moved except his head, which turned slowly.

"Some day I'll ride to the stars with you."

He said it as he watched the Six Flying Stars speed across the night sky—six glowing stars that moved in a direction opposite to the march of the other stars. Bright as Sirius seen from Earth, strung out one behind the other like jewels on a velvet string, they hurtled across the heavens.

They were the six ships on which the original Loire Valley Frenchmen had sailed out into space, seeking a home on a new planet. They had been put into an orbit around New Gaul and left there while their thirty thousand passengers had descended to the surface in chemical-fuel rockets. Mankind, once on the fair and fresh earth of the new planet, had never again ascended to re-visit the great ships.

For three hundred years the six ships had circled the planet known as New Gaul, nightly beacons and glowing reminders to Man that he was a stranger on this planet.

When the Earthmen landed on the new planet they had called the new land *Le Beau Pays*, or, as it was now pronounced, *L'Bawpfey*—The Beautiful Country. They had been delighted, entranced with the fresh new land. After the burned, war-racked Earth they had just left, it was like coming to Heaven.

They found two intelligent species living on the planet, and they found that the species lived in peace and that they had no conception of war or of poverty. And they were quite willing to receive the Terrans into their society.

Provided, that is, they became integrated, or—as they phrased it—natural. The Frenchmen from Earth had been given their choice. They were told:

"You can live with the people of the Beautiful Land on our terms—war with us, or leave to seek another planet."

The Terrans conferred. Half of them decided to stay; the other half decided to remain only long enough to mine uranium and other chemicals. Then they would voyage onwards.

But nobody from that group of Earthmen ever again stepped into the ferry-rockets and soared up to the six ion-beam ships circling about Le Beau Pays. All succumbed to the Philosophy of the Natural. Within a few generations a stranger landing upon the planet would not have known without previous information that the Terrans were not aboriginal.

He would have found three species. Two were warm-blooded egglayers who had evolved directly from reptiles without becoming mammals—the Ssassarors and the Amphibs. Somewhere in their dim past—like all happy nations, they had no history—they had set up their society and been very satisfied with it since.

It was a peaceful quiet world, largely peasant, where nobody had to scratch for a living and where a superb manipulation of biological forces ensured very long lives, no disease, and a social lubrication that left little to desire—from their viewpoint, anyway.

The government was, nominally, a monarchy. The Kings were elected by the people and were a different species than the group each ruled. Ssassaror ruled Human, and vice versa, each assisted by foster-brothers and sisters of the race over which they reigned. These were the so-called Dukes and Duchesses.

The Chamber of Deputies—*L'Syawp t' Tapfuti*—was half Human and half Ssassaror. The so-called Kings took turns presiding over the Chamber for forty day intervals. The Deputies were elected for ten-year terms by constituents who could not be deceived about their representatives' purposes because of the sensitive Skins which allowed them to determine their true feelings and worth.

In one custom alone did the ex-Terrans differ from their neighbors. This was in carrying arms. In the beginning, the Ssassaror had allowed the Men to wear their short rapiers, so they would feel safe even though in the midst of aliens.

As time went on, only the King's mucketeers—and members of the official underground—were allowed to carry épées. These men, it might be noticed, were the congenital adventurers, men who needed to swashbuckle and revel in the name of individualist.

Like the egg-stealers, they needed an institution in which they could work off anti-social steam.

From the beginning the Amphibians had been a little separate from the Ssassaror and when the Earthmen came they did not get any more neighborly. Nevertheless, they preserved excellent relations and they, too, participated in the Changeling-custom.

This Changeling-custom was another social device set up millennia ago to keep a mutual understanding between all species on the planet. It was a peculiar institution, one that the Earthmen had found hard to understand and ever more difficult to adopt. Nevertheless, once the Skins had been accepted they had changed their attitude, forgot their speculations about its origin and threw themselves into the custom of stealing babies—or eggs—from another race and raising the children as their own.

You rob my cradle; I'll rob yours. Such was their motto, and it worked.

A Guild of Egg Stealers was formed. The Human branch of it guaranteed, for a price, to bring you a Ssassaror child to replace the one that had been stolen from you. Or, if you lived on the sea-shore, and an Amphibian had crept into your nursery and taken your baby—always under two years old, according to the rules—then the Guildsman would bring you an Amphib or, perhaps, the child of a Human Changeling reared by the Seafolk.

You raised it and loved it as your own. How could you help loving it?

Your Skin told you that it was small and helpless and needed you and was, despite appearances, as Human as any of your babies. Nor did you need to worry about the one that had been abducted. It was getting just as good care as you were giving this one.

It had never occurred to anyone to quit the stealing and voluntary exchange of babies. Perhaps that was because it would strain even the loving nature of the Skin-wearers to give away their own flesh and blood. But once the transfer had taken place, they could adapt.

Or perhaps the custom was kept because tradition is stronger than law in a peasant-monarchy society and also because egg-and-baby stealing gave the more naturally aggressive and daring citizens a chance to work off anti-social behavior.

Nobody but a historian would have known, and there were no historians in The Beautiful Land.

Long ago the Ssassaror had discovered that if they lived meatless, they had a much easier time curbing their belligerency, obeying the Skins and remaining cooperative. So they induced the Earthmen to put a taboo on eating flesh. The only drawback to the meatless diet was that both Ssassaror and Man became as stunted in stature as they did in aggressiveness, the former so much so that they barely came to the chins of the Humans. These, in turn, would have seemed short to a Western European.

But Rastignac, an Earthman, and his good friend, Mapfarity, the Ssassaror Giant, became taboo-breakers when they were children and played together on the beach where they first ate seafood out of curiosity, then continued because they liked it. And due to their protein diet the Terran had grown well over six feet in height and the Ssassaror seemed to have touched off a rocket of expansion in his body with his protein-eating. Those Ssassarors who shared his guilt—became meat-eaters—became ostracized and eventually moved off to live by themselves. They were called Ssassaror-Giants and were pointed to as an object lesson to the young of the normal Ssassarors and Humans on the land.

If a stranger had landed shortly before Rastignac was born, however, he would have noticed that all was not as serene as it was supposed to be among the different species. The cause for the flaw in the former Eden might have puzzled him if he had not known the previous history of *L'Bawfey* and the fact that the situation had not changed for the worst until the introduction of Human Changelings among the Amphibians.

Then it had been that blood-drinking began among them, that Amphibians began seducing Humans to come live with them by their tales of easy immortality, and that they started the system of leaving savage little carnivores in the Human nurseries.

When the Land-dwellers protested, the Amphibs replied that these things were carried out by unnaturals or outlaws, and that the Sea-King could not be held responsible. Permission was given to Chalice those caught in such behavior.

Nevertheless, the suspicion remained that the Amphib monarch had, in accordance with age-old procedure, given his unofficial official blessing and that he was preparing even more disgusting and outrageous and unnatural moves. Through his control of the populace by the Master Skin, he would be able to do as he pleased with their minds.

It was the Skins that had made the universal peace possible on the planet of New Gaul. And it would be the custom of the Skins that would make possible the change from peace to conflict among the populace.

Through the artificial Skins that were put on all babies at birth—and which grew with them, attached to their body, feeding from their bloodstreams, their nervous systems—the Skins, controlled by a huge Master Skin that floated in a chemical vat in the palace of the rulers, fed, indoctrinated and attended day and night by a crew of the most brilliant scientists of the planet, gave the Kings complete control of the minds and emotions of the inhabitants of the planet.

Originally the rulers of New Gaul had desired only that the populace live in peace and enjoy the good things of their planet equally. But the change that had been coming gradually—the growth of conflict between the Kings of the different species for control of the whole populace—was beginning to be generally felt. Uneasiness, distrust of each other was growing among the people. Hence the legalizing of the Underground, the Philosophy of Violence by the government, an effort to control the revolt that was brewing.

Yet, the Land-dwellers had managed to take no action at all and to ignore the growing number of vicious acts.

But not all were content to drowse. One man was aroused. He was Rastignac.

They were Rastignac's hope, those Six Stars, the gods to which he prayed. When they passed quickly out of his sight he would continue his pacing, meditating for the twenty-thousandth time on a means for reaching one of those ships and using it to visit the stars. The end of his fantasies was always a curse because of the futility of such hopes. He was doomed! Mankind was doomed!

And it was all the more maddening because Man would not admit that he was through. Ended, that is, as a human being.

Man was changing into something not quite *homo sapiens*. It might be a desirable change, but it would mean the finish of his climb upwards. So it seemed to Rastignac. And he, being the man he was, had decided to do something about it even if it meant violence.

That was why he was now in the well-dungeon. He was an advocator of violence against the status quo.

Chapter II

There was another cell next to his. It was also at the bottom of a well and was separated from his by a thin wall of cement. A window had been set into it so that the prisoners could talk to each other. Rastignac did not care for the woman who had been let down into the adjoining cell, but she was somebody to talk to.

"Amphib-changelings" was the name given to those human beings who had been stolen from their cradles and raised among the non-humanoid Amphibians as their own. The girl in the adjoining cell, Lusine, was such a person. It was not her fault that she was a blood-drinking Amphib. Yet he could not help disliking her for what she had done and for the things she stood for.

She was in prison because she had been caught in the act of stealing a Man child from its cradle. This was no crime, but she had left in the cradle, under the covers, a savage and blood-thirsty little monster that had leaped up and slashed the throat of the unsuspecting baby's mother.

Her cell was lit by a cageful of glowworms. Rastignac, peering through the grille, could see her shadowy shape in the inner cell inside the wall. She rose langorously and stepped into the circle of dim orange light cast by the insects.

"B'zhu, m'fweh," she greeted him.

It annoyed him that she called him her brother, and it annoyed him even more to know that she knew it. It was true that she had some excuse for thus addressing him. She did resemble him. Like him, she had straight glossy blue-black hair, thick bracket-shaped eyebrows, brown eyes, a straight nose and a prominent chin. And where his build was superbly masculine, hers was magnificently feminine.

Nevertheless, this was not her reason for so speaking to him. She knew the disgust the Land-walker had for the Amphib-changeling, and she took a perverted delight in baiting him.

He was proud that he seldom allowed her to see that she annoyed him. *"B'zhu, fam tey zafeep,"* he said. "Good evening, woman of the Amphibians."

Mockingly she said, "Have you been watching the Six Flying Stars, Jean-Jacques?"

"*Vi.* I do so every time they come over."

"Why do you eat your heart out because you cannot fly up to them and then voyage among the stars on one of them?"

He refused to give her the satisfaction of knowing his real reason. He did not want her to realize how little he thought of Mankind and its chances for surviving—as humanity—upon the face of this planet, L'Bawpfey.

"I look at them because they remind me that Man was once captain of his soul."

"Then you admit that the Land-walker is weak?"

"I think he is on the way to becoming non-human, which is to say that he is weak, yes. But what I say about Landman goes for Seaman, too. You Changelings are becoming more Amphibian every day and less Human. Through the Skins the Amphibs are gradually changing you completely. Soon you will be completely sea-people."

She laughed scornfully, exposing perfect white teeth as she did so.

"The Sea will win out against the Land. It launches itself against the shore and shakes it with the crash of its body. It eats away the rock and the dirt and absorbs it into its own self. It can't be worn away nor caught and held in a net. It is elusive and all-powerful and never-tiring."

Lusine paused for breath. He said, "That is a very pretty analogy, but it doesn't apply. You Seafolk are as much flesh and blood as we Landfolk. What hurts us hurts you."

She put a hand around one bar. The glow-light fell upon it in such a way that it showed plainly the webbing of skin between her fingers. He glanced at it with a faint repulsion under which was a counter-current of attraction. This was the hand that had, indirectly, shed blood.

She glanced at him sidewise, challenged him in trembling tones. "You are not one to throw stones, Jean-Jacques. I have heard that you eat meat."

"Fish, not meat. That is part of my Philosophy of Violence," he retorted. "I maintain that one of the reasons man is losing his power and strength is that he has so long been upon a vegetable diet. He is as cowed and submissive as the grass-eating beast of the fields."

Lusine put her face against the bars.

"That is interesting," she said. "But how did you happen to begin eating fish? I thought we Amphibs alone did that."

What Lusine had just said angered him. He had no reply.

Rastignac knew he should not be talking to a Sea-changeling. They were glib and seductive and always searching for ways to twist your thoughts. But being Rastignac, he had to talk. Moreover, it was so difficult to find anybody who would listen to his ideas that he could not resist the temptation.

"I was given fish by the Ssassaror, Mapfarity, when I was a child. We lived along the sea-shore. Mapfarity was a child, too, and we played together. Don't eat fish!' my parents said. To me that meant 'Eat it!' So, despite my distaste at the idea, and my squeamish stomach, I did eat fish. And I liked it. And as I grew to manhood I adopted the Philosophy of Violence and I continued to eat fish although I am not a Changeling."

"What did your Skin do when it detected you?" Lusine asked. Her eyes were wide and luminous with wonder and a sort of glee as if she relished the confession of his sins. Also, he knew, she was taunting him about the futility of his ideas of violence so long as he was a prisoner of the Skin.

He frowned in annoyance at the reminder of the Skin. Much thought had he given, in a weak way, to the possibility of life without the Skin.

Ashamed now of his weak resistance to the Skin, he blustered a bit in front of the teasing Amphib girl.

"Mapfarity and I discovered something that most people don't know," he answered boastfully. "We found that if you can stand the shocks your Skin gives you when you do something wrong, the Skin gets tired and quits after a while. Of course your Skin recharges itself and the next time you eat fish it shocks you again. But after very many shocks it becomes accustomed, forgets its conditioning, and leaves you alone."

Lusine laughed and said in a low conspirational tone, "So your Ssassaror pal and you adopted the Philosophy of Violence because you remained fish and meat eaters?"

"Yes, we did. When Mapfarity reached puberty he became a Giant and went off to live in a castle in the forest. But we have remained friends through our connection in the underground."

"Your parents must have suspected that you were a fish eater when you first proposed your Philosophy of Violence?" she said.

"Suspicion isn't proof," he answered. "But I shouldn't be telling you all this, Lusine. I feel it is safe for me to do so only because you will never have a chance to tell on me. You will soon be taken to Chalice and there you will stay until you have been cured."

She shivered and said, "This Chalice? What is it?"

"It is a place far to the north where both Terrans and Ssassarors send their incorrigibles. It is an extinct volcano whose steep-sided interior makes an inescapable prison. There those who have persisted in unnatural behavior are given special treatment."

"They are bled?" she asked, her eyes widening as her tongue flicked over her lips again hungrily.

"No. A special breed of Skin is given them to wear. These Skins shock them more powerfully than the ordinary ones, and the shocks are associated with the habit they are trying to cure. The shocks effect a cure. Also, these special Skins are used to detect hidden unnatural emotions. They re-condition the deviate. The result is that when the Chaliced Man is judged able to go out and take his place in society again, he is thoroughly re-conditioned. Then his regular Skin is given back to him and it has no trouble keeping him in line from then on. The Chaliced Man is a very good citizen."

"And what if a revolter doesn't become Chaliced?"

"Then he stays in Chalice until he decides to become so."

Her voice rose sharply as she said, "But if I go there, and I am not fed the diet of the Amphibs, I will grow old and die!"

"No. The government will feed you the diet you need until you are re-conditioned. Except...." He paused.

"Except I won't get blood," she wailed. Then, realizing she was acting undignified before a Landman, she firmed her voice.

"The King of the Amphibians will not allow them to do this to me," she said. "When he hears of it he will demand my return. And if the King of Men refuses, my King will use violence to get me back."

Rastignac smiled and said, "I hope he does. Then perhaps my people will wake up and get rid of their Skins and make war upon your people."

"So that is what you Philosophers of Violence want, is it? Well, you will not get it. My father, the Amphib King, will not be so stupid as to declare a war."

"I suppose not," replied Rastignac. "He will send a band to rescue you. If they're caught they'll claim to be criminals and say they are *not* under the King's orders."

Lusine looked upwards to see if a guard was hanging over the well's mouth listening. Perceiving no one, she nodded and said, "You have guessed it correctly. And that is why we laugh so much at you stupid Humans. You know as well as we do what's going on, but you are afraid to tell us so. You keep clinging to the idea that your turn-the-other-cheek policy will soften us and insure peace."

"Not I," said Rastignac. "I know perfectly well there is only one solution to man's problems. That is—"

"That is Violence," she finished for him. "That is what you have been preaching. And that is why you are in this cell, waiting for trial."

"You don't understand," he said. "Men are not put into the Chalice for *proposing* new philosophies. As long as they behave naturally they may say what they wish. They may even petition the King that the new philosophy be made a law. The King passes it on to the Chamber of Deputies. They consider it and put it up to the people. If the people like it, it becomes a law. The only trouble with that procedure is that it may take ten years before the law is considered by the Chamber of Deputies."

"And in those ten years," she mocked him, "the Amphibs and the Amphibian-changelings will have won the planet."

"That is true," he said.

"The King of the Humans is a Ssassaror and the King of the Ssassaror is a Man," said Lusine. "Our King can't see any reason for changing the status quo. After all, it is the Ssassaror who are responsible for the Skins and for Man's position in the sentient society of this planet. Why should he be favorable to a policy of Violence? The Ssassarors loathe violence."

"And so you have preached Violence without waiting for it to become a law? And for that you are now in this cell?"

"Not exactly. The Ssassarors have long known that to suppress too much of Man's naturally belligerent nature only results in an explosion. So they have legalized illegality—up to a point. Thus the King officially made me the Chief of the Underground and gave me a state license to preach—but not practice—Violence. I am even allowed to advocate overthrow of the present system of government—as long as I take no action that is too productive of results.

"I am in jail now because the Minister of Ill-Will put me here. He had my Skin examined, and it was found to be 'unhealthy.' He thought I'd be better off locked up until I became 'healthy' again. But the King...."

Chapter III

Lusine's laughter was like the call of a silverbell bird. Whatever her unhuman appetites, she had a beautiful voice. She said, "How comical! And how do you, with your brave ideas, like being regarded as a harmless figure of fun, or as a sick man?"

"I like it as well as you would," he growled.

She gripped the bars of her window until the tendons on the back of her long thin hands stood out and the membranes between her fingers stretched like wind-blown tents. Face twisted, she spat at him, "Coward! Why don't you kill somebody and break out of this ridiculous mold—that Skin that the Ssassarors have poured you into?"

Rastignac was silent. That was a good question. Why didn't he? Killing was the logical result of his philosophy. But the Skin kept him docile. Yes, he could vaguely see that he had purposely shut his eyes to the destination towards which his ideas were slowly but inevitably traveling.

And there was another facet to the answer to her question—if he had to kill, he would not kill a Man. His philosophy was directed towards the Amphibians and the Sea-changelings.

He said, "Violence doesn't necessarily mean the shedding of blood, Lusine. My philosophy urges that we take a more vigorous action, that we overthrow some of the bio-social institutions which have imprisoned Man and stripped him of his dignity as an individual."

"Yes, I have heard that you want Man to stop wearing the Skin. That is what has horrified your people, isn't it?"

"Yes," he said. "And I understand it has had the same effect among the Amphibians."

She bridled, her brown eyes flashing in the feeble glowworms' light. "Why shouldn't it? What would we be without our Skins?"

"What, indeed?" he said, laughing derisively afterwards.

Earnestly she said, "You don't understand. We Amphibians—our Skins are not like yours. We do not wear them for the same reason you do. You are imprisoned by your Skins—they tell you how to feel, what to think.

Above all, they keep you from getting ideas about non-cooperation or non-integration with Nature as a whole.

"That, to us individualistic Amphibians, is false. The purpose of our Skins is to make sure that our King's subjects understand what he wants so that we may all act as one unit and thus further the progress of the Seafolk."

The first time Rastignac had heard this statement he had howled with laughter. Now, however, knowing that she could not see the fallacy, he did not try to argue the point. The Amphibs were, in their way, as hidebound—no pun intended—as the Land-walkers.

"Look, Lusine," he said, "there are only three places where a Man may take off his Skin. One is in his own home, when he may hang it upon the halltree. Two is when he is, like us, in jail and therefore may not harm anybody. The third is when a man is King. Now you and I have been without our Skins for a week. We have gone longer without them than anybody, except the King. Tell me true, don't you feel free for the first time in your life?

"Don't you feel as if you belong to nobody but yourself, that you are accountable to no one but yourself, and that you love that feeling? And don't you dread the day we will be let out of prison and made to wear our Skins again? That day which, curiously enough, will be the very day that we will lose our freedom."

Lusine looked as if she didn't know what he was talking about.

"You'll see what I mean when we are freed and the Skins are put back upon us," he said. Immediately after, he was embarrassed. He remembered that she would go to the Chalice where one of the heavy and powerful Skins used for unnaturals would be fastened to her shoulders.

Lusine did not notice. She was considering the last but most telling point in her argument "You cannot win against us," she said, watching him narrowly for the effect of her words. "We have a weapon that is irresistible. We have immortality."

His face did not lose its imperturbability.

She continued, "And what is more, we can give immortality to anyone who casts off his Skin and adopts ours. Don't think that your people don't know this. For instance, during the last year more than two thousand Humans living along the beaches deserted and went over to us, the Amphibs."

He was a little shocked to hear this, but he did not doubt her. He remembered the mysterious case of the schooner *Le Pauvre Pierre* which had been found drifting and crewless, and he remembered a conversation he had had with a fisherman in his home port of Marrec.

He put his hands behind his back and began pacing. Lusine continued staring at him through the bars. Despite the fact that her face was in the shadows, he could see—or feel—her smile. He had humiliated her, but she had won in the end.

Rastignac quit his limited roving and called up to the guard.

"Shoo l'footyay, kal u ay tee?"

The guard leaned over the grille. His large hat with its tall wings sticking from the peak was green in the daytime. But now, illuminated only by a far off torchlight and by a glowworm coiled around the band, it was black.

"Ah, shoo Zhaw-Zhawk W'stenyek," he said, loudly. "What time is it? What do you care what time it is?" And he concluded with the stock phrase of the jailer, unchanged through millenia and over light-years. "You're not going any place, are you?"

Rastignac threw his head back to howl at the guard but stopped to wince at the sudden pain in his neck. After uttering, *"Sek Ploo!"* and *"S'pweestee!"* both of which were close enough to the old Terran French so that a language specialist might have recognized them, he said, more calmly, "If you would let me out on the ground, *monsieur le foutriquet*, and give me a good épée, I would show you where I am going. Or, at least, where my sword is going. I am thinking of a nice sheath for it."

Tonight he had a special reason for keeping the attention of the King's mucketeer directed towards himself. So, when the guard grew tired of returning insults—mainly because his limited imagination could invent no new ones—Rastignac began telling jokes. They were broad and aimed at the mucketeer's narrow intellect.

"Then," said Rastignac, "there was the itinerant salesman whose *s'fel* threw a shoe. He knocked on the door of the hut of the nearest peasant and said...." What was said by the salesman was never known.

A strangled gasp had come from above.

Chapter IV

Rastignac saw something enormous blot out the smaller shadow of the guard. Then both figures disappeared. A moment later a silhouette cut across the lines of the grille. Unoiled hinges screeched; the bars lifted. A rope uncoiled from above to fall at Rastignac's feet. He seized it and felt himself being drawn powerfully upwards.

When he came over the edge of the well, he saw that his rescuer was a giant Ssassaror. The light from the glowworm on the guard's hat lit up feebly his face, which was orthagnathous and had quite humanoid eyes and lips. Large canine teeth stuck out from the mouth, and its huge ears were tipped with feathery tufts. The forehead down to the eyebrows looked as if it needed a shave, but Rastignac knew that more light would show the blue-black shade came from many small feathers, not stubbled hair.

"Mapfarity!" Rastignac said. "It's good to see you after all these years!"

The Ssassaror giant put his hand on his friend's shoulder. Clenched, it was almost as big as Rastignac's head. He spoke with a voice like a lion coughing at the bottom of a deep well.

"It is good to see you again, my friend."

"What are you doing here?" said Rastignac, tears running down his face as he stroked the great fingers on his shoulder.

Mapfarity's huge ears quivered like the wings of a bat tied to a rock and unable to fly off. The tufts of feathers at their ends grew stiff and suddenly crackled with tiny sparks.

The electrical display was his equivalent of the human's weeping. Both creatures discharged emotion; their bodies chose different avenues and manifestations. Nevertheless, the sight of the other's joy affected each deeply.

"I have come to rescue you," said Mapfarity. "I caught Archambaud here,"—he indicated the other man—"stealing eggs from my golden goose. And...."

Raoul Archambaud—pronounced Wawl Shebvo—interrupted excitedly, "I showed him my license to steal eggs from Giants who were raising counterfeit geese, but he was going to lock me up anyway. He was going to take my Skin off and feed me on meat...."

"Meat!" said Rastignac, astonished and revolted despite himself. "Mapfarity, what have you been doing in that castle of yours?"

Mapfarity lowered his voice to match the distant roar of a cataract. "I haven't been very active these last few years," he said, "because I am so big that it hurts my feet if I walk very much. So I've had much time to think. And I, being logical, decided that the next step after eating fish was eating meat. It couldn't make me any larger. So, I ate meat. And while doing so, I cam to the same conclusion that you, apparently, have done independently. Tha is, the Philosophy of...."

"Of Violence," interrupted Archambaud. "Ah, Jean-Jacques, there mus be some mystic bond that brings two Humans of such different backgrounds as yours and the Ssassaror together, giving you both the same philosophy. When I explained what you had been doing and that you were in jail because you had advocated getting rid of the Skins, Mapfarity petitioned...."

"The King to make an official jail-break," said Mapfarity with an impatient glance at the rolypoly egg-stealer. "And...."

"The King agreed," broke in Archambaud, "provided Mapfarity would turn in his counterfeit goose and provided you would agree to say no more about abandoning Skins, but...."

The Giant's basso profundo-redundo pushed the egg-stealer's high pitch aside. "If this squeaker will quit interrupting, perhaps we can get on with the rescue. We'll talk later, if you don't mind."

At that moment Lusine's voice floated up from the bottom of her cell. "Jean-Jacques, my love, my brave, my own, would you abandon me to the Chalice? Please take me with you! You will need somebody to hide you when the Minister of Ill-Will sends his mucketeers after you. I can hide you where no one will ever find you." Her voice was mocking, but there was an undercurrent of anxiety to it.

Mapfarity muttered, "She will hide us, yes, at the bottom of a sea-cave where we will eat strange food and suffer a change. Never...."

"Trust an Amphib," finished Archambaud for him.

Mapfarity forgot to whisper. "*Bey-t'cul, vu nu fez yey! Fe'm sa!*" he roared.

A shocked hush covered the courtyard. Only Mapfarity's wrathful breathing could be heard. Then, disembodied, Lusine's voice floated from the well.

"Jean-Jacques, do not forget that I am the foster-daughter of the King of the Amphibians! If you were to take me with you, I could assure you of safety and a warm welcome in the halls of the Sea-King's Palace!"

"Pah!" said Mapfarity. "That web-footed witch!"

Rastignac did not reply to her. He took the broad silk belt and the sheathed épée from Archambaud and buckled them around his waist. Mapfarity handed him a mucketeer's hat; he clapped that on firmly. Last of all, he took the Skin that the fat egg-stealer had been holding out to him.

For the first time he hesitated. It was his Skin, the one he had been wearing since he was six. It had grown with him, fed off his blood for twenty-two years, clung to him as clothing, censor, and castigator, and parted from him only when he was inside the walls of his own house, went swimming, or, as during the last seven days, when he laid in jail.

A week ago, after they had removed his second Skin, he had felt naked and helpless and cut off from his fellow creatures. But that was a week ago. Since then, as he had remarked to Lusine, he had experienced the birth of a strange feeling. It was, at first, frightening. It made him cling to the bars as if they were the only stable thing in the center of a whirling universe.

Later, when that first giddiness had passed, it was succeeded by another intoxication—the joy of being an individual, the knowledge that he was separate, not a part of a multitude. Without the Skin he could think as he pleased. He did not have a censor.

Now, he was on level ground again, out of the cell. But as soon as he had put that prison-shaft behind him he was faced with the old second Skin.

Archambaud held it out like a cloak in his hands. It looked much like a ragged garment. It was pale and limp and roughly rectangular with four extensions at each corner. When Rastignac put it on his back, it would sink four tiny hollow teeth into his veins and the suckers on the inner surface of its flat body would cling to him. Its long upper extensions would wrap themselves around his shoulders and over his chest; the lower, around his loins and thighs. Soon it would lose its paleness and flaccidity, become pink and slightly convex, pulsing with Rastignac's blood.

Chapter V

Rastignac hesitated for a few seconds. Then he allowed the habit of a lifetime to take over. Sighing, he turned his back. In a moment he felt the cold flesh descend over his shoulders and the little bite of the four teeth as they attached the Skin to his shoulders. Then, as his blood poured into the creature he felt it grow warm and strong. It spread out and followed the passages it had long ago been conditioned to follow, wrapped him warmly and lovingly and comfortably. And he knew, though he couldn't feel it, that it was pushing nerves into the grooves along the teeth. Nerves to connect with his.

A minute later he experienced the first of the expected *rapport*. It was nothing that you could put a mental finger on. It was just a diffused tingling and then the sudden consciousness of how the others around him *felt*.

They were ghosts in the background of his mind. Yet, pale and ectoplasmic as they were, they were easily identifiable. Mapfarity loomed above the others, a transparent Colossus radiating streamers of confidence in his clumsy strength. A meat-eater, uncertain about the future, with a hope and trust in Rastignac to show him the right way. And with a strong current of anger against the conqueror who had inflicted the Skin upon him.

Archambaud was a shorter phantom, rolypoly even in his psychic manifestations, emitting bursts of impatience because other people did not talk fast enough to suit him, his mind leaping on ahead of their tongues, his fingers wriggling to wrap themselves around something valuable—preferably the eggs of the golden goose—and a general eagerness to be up and about and onwards. He was one round fidget on two legs, yet a good man for any project requiring action.

Faintly, Rastignac detected the slumbering guard as if he were the tendrils of some plant at the sea-bottom, floating in the green twilight, at peace and unconscious.

And even more faintly he felt Lusine's presence, shielded by the walls of the shaft. Hers was a pale and light hand, one whose fingers tapped a barely heard code of impotent rage and voiceless screaming fear. Yet beneath that anguish was a base of confidence and mockery at others. She might be temporarily upset, but when the chance came for her to do something she would seize it with every ability at her command.

Another radiation dipped into the general picture and out. A wild glowworm had swooped over them and disturbed the smooth reflection built up by the Skins.

This was the way the Skins worked. They penetrated into you and found out what you were feeling and emoting, and then they broadcast it to other closeby Skins, which then projected their hosts' psychosomatic responses. The whole was then integrated so that each Skin-wearer could detect the group-feeling and at the same time, though in a much duller manner, the feeling of the individuals of the *gestalt*.

That wasn't the only function of the Skin. The parasite, created in the bio-factories, had several other social and biological uses.

Rastignac almost fell into a reverie at that point. It was nothing unusual. The effect of the Skins was a slowing-down one. The wearer thought more slowly, acted more leisurely, and was much more contented.

But now, by a deliberate wrenching of himself from the feeling-pattern, Rastignac woke up. There were things to do, and standing around and drinking in the lotus of the group-rapport was not one of them.

He gestured at the prostrate form of the mucketeer. "You didn't hurt him?"

The Ssassaror rumbled, "No. I scratched him with a little venom of the dream-snake. He will sleep for an hour or so. Besides, I would not be allowed to hurt him. You forget that all this is carefully staged by the King's Official Jail-breaker."

"*Me'dt!*" swore Rastignac.

Alarmed, Archambaud said, "What's the matter, Jean-Jacques?"

"Can't we do anything on our own? Must the King meddle in everything?"

"You wouldn't want us to take a chance and have to shed *blood,* would you?" breathed Archambaud.

"What are you carrying those swords for? As a decoration?" Rastignac snarled.

"Seelahs, m'fweh," warned Mapfarity. "If you alarm the other guards, you will embarrass them. They will be forced to do their duty and recapture you. And the Jail-breaker would be reprimanded because he had fallen down on his job. He might even get a demotion."

Rastignac was so upset that his Skin, reacting to the negative fields racing over the Skin and the hormone imbalance of his blood, writhed away from his back.

"What are we, a bunch children playing war?"

Mapfarity growled, "We are all God's children, and we mustn't hurt anyone if we can help it."

"Mapfarity, you eat meat!"

"Voo zavf w'zaw m'fweh," admitted the Giant. "But it is the flesh of unintelligent creatures. I have not yet shed the blood of any being that can talk with the tongue of Man."

Rastignac snorted and said, "If you stick with me you will some day do that, *m'fweh* Mapfarity. There is no other course. It is inevitable."

"Nature spare me the day! But if it comes it will find Mapfarity unafraid. They do not call me Giant for nothing."

Rastignac sighed and walked ahead. Sometimes he wondered if the members of his underground—or anybody else for that matter—ever realized the grim conclusions formed by the Philosophy of Violence.

The Amphibians, he was sure, did. And they were doing something positive about it. But it was the Amphibians who had driven Rastignac to adopt a Philosophy of Violence.

"Law," he said again. "Let's go."

The three of them walked out of the huge courtyard and through the open gate. Nearby stood a short man whose Skin gleamed black-red in the light shed by the two glowworms attached to his shoulders. The Skin was oversized and hung to the ground.

The King's man, however, did not think he was a comic figure. He sputtered, and the red of his face matched the color of the skin on his back.

"You took long enough," he said accusingly and then, when Rastignac opened his mouth to protest, the Jail-breaker said, "Never mind, never mind. *Sa n'apawt*. The thing is that we get you away fast. The Minister of

Ill-Will has doubtless by now received word that an official jail-break is planned for tonight. He will send a company of his mucketeers to intercept you. By coming in advance of the appointed time we shall have time to escape before the official rescue party arrives."

"How much time do we have?" asked Rastignac.

The King's man said, "Let's see. After I escort you through the rooms of the Duke, the King's foster-brother—he is most favorable to the Violent Philosophy, you know, and has petitioned the King to become your official patron, which petition will be considered at the next meeting of the Chamber of Deputies in three months—let's see, where was I? Ah, yes, I escort you through the rooms of the King's brother. You will be disguised as His Majesty's mucketeers, ostensibly looking for the escaped prisoners. From the rooms of the Duke you will be let out through a small door in the wall of the palace itself. A car will be waiting.

"From then on it will be up to you. I suggest, however, that you make a dash for Mapfarity's castle. Follow the *Rue des Nues*; that is your best chance. The mucketeers have been pulled off that boulevard. However, it is possible that Auverpin, the Ill-Will Minister, may see that order and will rescind it, realizing what it means. If he does, I suppose I will see you back in your cell, Rastignac."

He bowed to the Ssassaror and Archambaud and said, "And you two gentlemen will then be with him."

"And then what?" rumbled Mapfarity.

"According to the law, you will be allowed one more jail-break. Any more after that will, of course, be illegal. That is, unthinkable."

Rastignac unsheathed his épée and slashed it at the air. "Let the mucketeers stand in my way," he said fiercely. "I will cut them down with this!"

The Jail-breaker staggered back, hands outthrust.

"Please, Monsieur Rastignac! Please! Don't even talk about it! You know that your philosophy is, as yet, illegal. The shedding of blood is an act that will be regarded with horror throughout the sentient planet. People would think you are an Amphibian!"

"The Amphibians know what they're doing far better than we do," answered Rastignac. "Why do you think they're winning against us Humans?"

Suddenly, before anybody could answer, the sound of blaring horns came from somewhere on the ramparts. Shouts went up; drums began to beat, calling the mucketeers to alert.

And above it all came the roar of a giant Ssassaror voice: *"An Earthship has landed in the sea! And the pilot of the ship is in the hands of the Amphibians!"*

As the meaning of the words seeped into Rastignac's consciousness he made a sudden violent movement—and began to tear the Skin from his body!

Chapter VI

Rastignac ran down the steps, out into the courtyard. He seized the Jailbreaker's arm and demanded the key to the grilles. Dazed, the white-faced official meekly and silently handed it to him. Without his Skin Rastignac was no longer fearfully inhibited. If you were forceful enough and did not behave according to the normal pattern you could get just about anything you wanted. The average Man or Ssassaror did not know how to react to his violence. By the time they had recovered from their confusion he could be miles away.

Such a thought flashed through his head as he ran towards the prison wells. At the same time he heard the horn-blasts of the king's mucketeers and knew that he shortly would have a different type of Man to deal with. The mucketeers, closest approach to soldiers in this pacifistic land, wore Skins that conditioned them to be more belligerent than the common citizen. They carried épées and, while it was true that their points were dull and their wielders had never engaged in serious swordsmanship, the mucketeers could be dangerous from a viewpoint of numbers alone.

Mapfarity bellowed, "Jean-Jacques, what are you doing?"

He called back over his shoulder, "I'm taking Lusine with us! She can help us get the Earthman from the Amphibians!"

The Giant lumbered up behind him, threw a rope down to the eager hands of Lusine and pulled her up without effort to the top of the well. A second later, Rastignac leaped upon Mapfarity's back, dug his hands under the upper fringe of the huge Skin and, ignoring its electrical blasts, ripped downwards.

Mapfarity cried out with shock and surprise as his skin flopped on the stones like a devilfish on dry land.

Archambaud ran up then and, without bothering to explain, the Ssassaror and the Man seized him and peeled off *his* artificial hide.

"Now we're all free men!" panted Rastignac. "And the mucketeers have no way of locating us if we hide, nor can they punish us with shocks."

He put the Giant on his right side, Lusine on his left, and the egg-stealer behind him. He removed the Jail-breaker's rapier from his sheath. The official was too astonished to protest.

"Law, m'zawfa!" cried Rastignac, parodying in his grotesque French the old Gallic war cry of *"Allons, mes enfants!"*

The King's official came to life and screamed orders at the group of mucketeers who had poured into the courtyard. They halted in confusion. They could not hear him above the roar of horns and thunder of drums and the people sticking their heads out of windows and shouting.

Rastignac scooped up with his épée one of the abandoned Skins flopping on the floor and threw it at the foremost guard. It descended upon the man's head, knocking off his hat and wrapping itself around the head and shoulders. The guard dropped his sword and staggered backwards into the group. At the same time the escapees charged and bowled over their feeble opposition.

It was here that Rastignac drew first blood. The tip of his épée drove past a bewildered mucketeer's blade and entered the fellow's throat just below the chin. It did not penetrate very far because of the dullness of the point. Nevertheless, when Rastignac withdrew his sword he saw blood spurt.

It was the first flower of violence, this scarlet blossom set against the whiteness of a Man's skin.

It would, if he had worn his Skin, have sickened him. Now, he exulted with a shout of triumph.

Lusine swooped up from behind him, bent over the fallen man. Her fingers dipped into the blood and went to her mouth. Greedily, she sucked her fingers.

Rastignac struck her cheek hard with the flat of his hand. She staggered back, her eyes narrow, but she laughed.

The next moments were busy as they entered the castle, knocked down two mucketeers who tried to prevent their passage to the Duke's rooms, then filed across the long suite.

The Duke rose from his writing-desk to greet them. Rastignac, determined to sever all ties and impress the government with the fact that he meant a real violence, snarled at his benefactor, *"Va t'feh fout!"*

The Duke was disconcerted at this harsh command, so obviously impossible to carry out. He blinked and said nothing. The escapees hurried past him to the door that gave exit to the outside. They pushed it open and stepped out into the car that waited for them. A chauffeur leaned against its thin wooden body.

Mapfarity pushed him aside and climbed in. The others followed. Rastignac was the last to get in. He examined in a glance the vehicle they were supposed to make their flight in.

It was as good a car as you could find in the realm. A Renault of the large class, it had a long boat-shaped scarlet body. There wasn't a scratch on it. It had seats for six. And that it had the power to outrun most anything was indicated by the two extra pairs of legs sticking out from the bottom. There were twelve pairs of legs, equine in form and shod with the best steel. It was the kind of vehicle you wanted when you might have to take off across the country. Wheeled cars could go faster on the highway, but this Renault would not be daunted by water, plowed fields, or steep hillsides.

Rastignac climbed into the driver's seat, seized the wheel and pressed his foot down on the accelerator. The nerve-spot beneath the pedal sent a message to the muscles hidden beneath the hood and the legs projecting from the body. The Renault lurched forward, steadied, and began to pick up speed. It entered a broad paved highway. Hooves drummed; sparks shot out from the steel shoes.

Rastignac guided the brainless, blind creature concealed within the body. He was helped by the somatically-generated radar it employed to steer it past obstacles. When he came to the *Rue des Nues*, he slowed it down to a trot. There was no use tiring it out. Halfway up the gentle slope of the boulevard, however, a Ford galloped out from a side-street. Its seats bristled with tall peaked hats with outspread glowworm wings and with drawn épées.

Rastignac shoved the accelerator to the floor. The Renault broke into a gallop. The Ford turned so that it would present its broad side. As there was a fencework of tall shrubbery growing along the boulevard, the Ford was thus able to block most of the passage.

But, just before his vehicle reached the Ford, Rastignac pressed the Jump button. Few cars had this; only sportsmen or the royalty could afford to have such a neural circuit installed. And it did not allow for gradations in leaping. It was an all-or-none reaction; the legs spurned the ground in perfect unison and with every bit of the power in them. There was no holding back.

The nose lifted, the Renault soared into the air. There was a shout, a slight swaying as the trailing hooves struck the heads of mucketeers who had been stupid enough not to duck, and the vehicle landed with a screeching lurch, upright, on the other side of the Ford. Nor did it pause.

Half an hour later Rastignac reined in the car under a large tree whose shadow protected them. "We're well out in the country," he said.

"What do we do now?" asked impatient Archambaud.

"First we must know more about this Earthman," Rastignac answered. "Then we can decide."

Chapter VII

Dawn broke through night's guard and spilled a crimson swath on the hills to the East, and the Six Flying Stars faded from sight like a necklace of glowing jewels dipped into an ink bottle.

Rastignac halted the weary Renault on the top of a hill, looked down over the landscape spread out for miles below him. Mapfarity's castle—a tall rose-colored tower of flying buttresses—flashed in the rising sun. It stood on another hill by the sea shore. The country around was a madman's dream of color. Yet to Rastignac every hue sickened the eye. That bright green, for instance, was poisonous; that flaming scarlet was bloody; that pale yellow, rheumy; that velvet black, funeral; that pure white, maggotty.

"Rastignac!" It was Mapfarity's bass, strumming irritation deep in his chest.

"What?"

"What do we do now?"

Jean-Jacques was silent. Archambaud spoke plaintively.

"I'm not used to going without my Skin. There are things I miss. For one thing, I don't know what you're thinking, Jean-Jacques. I don't know whether you're angry at me or love me or are indifferent to me. I don't know where other people *are*. I don't feel the joy of the little animals playing, the freedom of the flight of birds, the ghostly plucking of the growing grass, the sweet stab of the mating lust of the wild-horned apigator, the humming of bees working to build a hive, and the sleepy stupid arrogance of the giant cabbage-eating *deuxnez*. I can feel nothing without the Skin I have worn so long. I feel alone."

Rastignac replied, "You are not alone. I am with you."

Lusine spoke in a low voice, her large brown eyes upon his.

"I, too, feel alone. My Skin is gone, the Skin by which I knew how to act according to the wisdom of my father, the Amphib King. Now that it is gone and I cannot hear his voice through the vibrating tympanum, I do not know what to do."

"At present," replied Rastignac, "you will do as I tell you."

Mapfarity repeated, "What now?"

Rastignac became brisk. He said, "We go to your castle, Giant. We use your smithy to put sharp points on our swords, points to slide through a man's body from front to back. Don't pale! That is what we must do. And then we pick up your goose that lays the golden eggs, for we must have money if we are to act efficiently. After that, we buy—or steal—a boat and we go to wherever the Earthman is held captive. And we rescue him."

"And then?" said Lusine, her eyes shining with emotion.

"What you do then will be up to you. But I am going to leave this planet and voyage with the Earthman to other worlds."

Silence. Then Mapfarity said, "Why leave here?"

"Because there is no hope for this land. Nobody will give up his Skin. *Le Beau Pays* is doomed to a lotus-life. And that is not for me."

Archambaud jerked a thumb at the Amphib girl. "What about her people?"

"They may win, the water-people. What's the difference? It will be just the exchange of one Skin for another. Before I heard of the landing of the Earthman I was going to fight no matter what the cost to me or inevitable defeat. But not now."

Mapfarity's rumble was angry. "Ah, Jean-Jacques, this is not my comrade talking. Are you sure you haven't swallowed your Skin? You talk as if you were inside-out. What is the matter with your brain? Can't you see that it will indeed make a difference if the Amphibs get the upper hand? Can't you see *who* is making the Amphibs behave the way they have been?"

Rastignac urged the Renault towards the rose-colored lacy castle high upon a hill. The vehicle trotted tiredly along the rough and narrow forest path.

"What do you mean?" he said.

"I mean the Amphibs got along fine with the Ssassaror until a new element entered their lives—the Earthmen. Then the antagonising began. What is this new element? It's the Changelings—the mixture of Earthmen and Amphibs or Ssassaror and Terran. Add it up. Turn it around. Look at it from any angle. It is the Changelings who are behind this restlessness—the Human element.

"Another thing. The Amphibs have always had Skins different from ours. Our factories create our Skins to set up an affinity and communication between their wearers and all of Nature. They are designed to make it easier for every Man to love his neighbor.

"Now, the strange thing about the Amphibs' Skin is that they, too, were once designed to do such things. But in the past thirty or forty years new Skins have been created for one primary purpose—to establish a communication between the Sea-King and his subjects. Not only that, the Skins can be operated at long distances so that the King may punish any disobedient subject. And they are set so that they establish affinity only among the Waterfolk, not between them and all of Nature."

"I had gathered some of that during my conversations with Lusine," said Rastignac. "But I did not know it had gone to such lengths."

"Yes, and you may safely bet that the Changelings are behind it."

"Then it is the human element that is corrupting?"

"What else?"

Rastignac said, "Lusine, what do you say to this?"

"I think it is best that you leave this world. Or else turn Changeling-Amphib."

"Why should I join you Amphibians?"

"A man like you could become a Sea-King."

"And drink blood?"

"I would rather drink blood than mate with a Man. Almost, that is. But I would make an exception with you, Jean-Jacques."

If it had been a Land-woman who made such a blunt proposal he would have listened with equanimity. There was no modesty, false or otherwise in the country of the Skin-wearers. But to hear such a thing from a woman whose mouth had drunk the blood of a living man filled him with disgust.

Yet, he had to admit Lusine was beautiful. If she had not been a blood-drinker....

Though he lacked his receptive Skin, Mapfarity seemed to sense Rastignac's emotions. He said, "You must not blame her too much, Jean-Jacques. Sea-changelings are conditioned from babyhood to love blood. And for a very definite purpose, too, unnatural though it is. When the time comes for hordes of Changelings to sweep out of the sea and overwhelm the Landfolk, they will have no compunctions about cutting the throats of their fellow-creatures."

Lusine laughed. The rest of them shifted uneasily but did not comment. Rastignac changed the subject.

"How did you find out about the Earthman, Mapfarity?" he said.

The Ssassaror smiled. Two long yellow canines shone wetly; the nose, which had nostrils set in the sides, gaped open; blue sparks shot out from it; at the same time the feathered tufts on the ends of the elephantine ears stiffened and crackled with red-and-blue sparks.

"I have been doing something besides breeding geese to lay golden eggs," he said. "I have set traps for Waterfolk, and I have caught two. These I caged in a dungeon in my castle, and I experimented with them. I removed their Skins and put them on me, and I found out many interesting facts."

He leered at Lusine, who was no longer laughing, and he said, "For instance, I discovered that the Sea-King can locate, talk to, and punish any of his subjects anywhere in the sea or along the coast. He has booster Skins planted all over his realm so that any message he sends will reach the receiver, no matter how far away he is. Moreover, he has conditioned each and every Skin so that, by uttering a certain code-word to which only one particular Skin will respond, he may stimulate it to shock or even to kill its carrier."

Mapfarity continued, "I analyzed those two Skins in my lab and then, using them as models, made a number of duplicates in my fleshforge. They lacked only the nerves that would enable the Sea-King to shock us."

Rastignac smiled his appreciation of this coup. Mapfarity's ears crackled blue sparks of joy, his equivalent of blushing.

"Ah, then you have doubtless listened in to many broadcasts. And you know where the Earthman is located?"

"Yes," said the Giant. "He is in the palace of the Amphib King, upon the island of Kataproimnoin. That is only thirty miles out to the sea."

Rastignac did not know what he would do, but he had two advantages in the Amphibs' Skins and in Lusine. And he burned to get off this doomed planet, this land of men too sunk in false happiness, sloth, and stupidity to see that soon death would come from the water.

He had two possible avenues of escape. One was to use the newly arrived Earthman's knowledge so that the fuels necessary to propel the ferry-rockets could be manufactured. The rockets themselves still stood in a museum. Rastignac had not planned to use them because neither he nor any one else on this planet knew how to make fuel for them. Such secrets had long ago been forgotten.

But now that science was available through the newcomer from Earth, the rockets could be equipped and taken up to one of the Six Flying Stars. The Earthman could study the rocket, determine what was needed in the way of supplies, then it could be outfitted for the long voyage.

An alternative was the Terran's vessel. Perhaps he might invite him to come along in it....

The huge gateway to Mapfarity's castle interrupted his thoughts.

Chapter VIII

He halted the Renault, told Archambaud to find the Giant's servant and have him feed their vehicle, rub its legs down with liniment, and examine the hooves for defective shoes.

Archambaud was glad to look up Mapfabvisheen, the Giant's servant, because he had not seen him for a long time. The little Ssassaror had been an active member of the Egg-stealer's Guild until the night three years ago when he had tried to creep into Mapfarity's strongroom. The crafty guildsman had avoided the Giant's traps and there found the two geese squatting upon their bed of minerals.

These fabulous geese made no sound when he picked them up with lead-lined gloves and put them in his bag, also lined with lead-leaf. They were not even aware of him. Laboratory-bred, retort-shaped, their protoplasm a blend of silicon-carbon, unconscious even that they lived, they munched upon lead and other elements, ruminated, gestated, transmuted, and every month, regular as the clockwork march of stars or whirl of electrons, each laid an octagonal egg of pure gold.

Mapfabvisheen had trodden softly from the strongroom and thought himself safe. And then, amazingly, frighteningly, and totally unethically, from his viewpoint, the geese had begun honking loudly!

He had run, but not fast enough. The Giant had come stumbling from his bed in response to the wild clamor and had caught him. And, according to the contract drawn up between the Guild of Egg-stealers and the League of Giants, a guildsman seized within the precincts of a castle must serve the goose's owner for two years. Mapfabvisheen had been greedy; he had tried to take both geese. Therefore, he must wait upon the Giant for a double term.

Afterwards, he found out how he'd been trapped. The egglayers themselves hadn't been honking. Mouthless, they were utterly incapable of that. Mapfarity had fastened a so-called "goose-tracker" to the strong-room's doorway. This device clicked loudly whenever a goose was nearby. It could smell out one even through a lead-leaf-lined bag. When Mapfabvisheen passed underneath it, its clicks woke up a small Skin beside it. The Skin, mostly lung-sac and voice organs, honked its warning. And the dwarf, Mapfabvisheen, began his servitude to the Giant, Mapfarity.

Rastignac knew the story. He also knew that Mapfarity had infected the fellow with the philosophy of Violence and that he was now a good member of his Underground. He was eager to tell him his servitor days were over, that he could now take his place in their band as an equal. Subject, of course, to Rastignac's order.

Mapfabvisheen was stretched out upon the floor and snoring a sour breath. A grey-haired man was slumped on a nearby table. His head, turned to one side, exhibited the same slack-jawed look that the Ssassaror's had, and he flung the ill-smelling gauntlet of his breath at the visitors. He held an empty bottle in one loose hand. Two other bottles lay on the stone floor, one shattered.

Besides the bottles lay the men's Skins. Rastignac wondered why they had not crawled to the halltree and hung themselves up.

"What ails them? What is that smell?" said Mapfarity.

"I don't know," replied Archambaud, "but I know the visitor. He is Father Jules, priest of the Guild of Egg-stealers."

Rastignac raised his queer, bracket-shaped eyebrows, picked up a bottle in which there remained a slight residue, and drank.

"Mon Dieu, it is the sacrament wine!" he cried.

Mapfarity said, "Why would they be drinking that?"

"I don't know. Wake Mapfabvisheen up, but let the good father sleep. He seems tired after his spiritual labors and doubtless deserves a rest."

Doused with a bucket of cold water the little Ssassaror staggered to his feet. Seeing Archambaud, he embraced him. "Ah, Archambaud, old baby-abductor, my sweet goose-bagger, my ears tingle to see you again!"

They did. Red and blue sparks flew off his ear-feathers.

"What is the meaning of this?" sternly interrupted Mapfarity. He pointed at the dirt swept into the corners.

Mapfabvisheen drew himself up to his full dignity, which wasn't much. "Good Father Jules was making his circuits," he said. "You know he travels around the country and hears confession and sings Mass for us poor egg-stealers who have been unlucky enough to fall into the clutches of some rich and greedy and anti-social Giant who is too stingy to hire servants, but captures them instead, and who won't allow us to leave the premises until our servitude is over...."

"Cut it!" thundered Mapfarity. "I can't stand around all day, listening to the likes of you. My feet hurt too much. Anyway, you know I've allowed you to go into town every week-end. Why don't you see a priest then?"

Mapfabvisheen said, "You know very well the closest town is ten kilometers away and it's full of Pantheists. There's not a priest to be found there."

Rastignac groaned inwardly. Always it was thus. You could never hurry these people or get them to regard anything seriously.

Take the case they were wasting their breath on now. Everybody knew the Church had been outlawed a long time ago because it opposed the use of the Skins and certain other practices that went along with it. So, no sooner had that been done than the Ssassarors, anxious to establish their check-and-balance system, had made arrangements through the Minister of Ill-Will to give the Church unofficial legal recognizance.

Then, though the aborigines had belonged to that pantheistical organization known as the Sons of Good And Old Mother Nature, they had all joined the Church of the Terrans. They operated under the theory that the best way to make an institution innocuous was for everybody to sign up for it. Never persecute. That makes it thrive.

Much to the Church's chagrin, the theory worked. How can you fight an enemy who insists on joining you and who will also agree to everything you teach him and then still worship at the other service? Supposedly driven underground, the Church counted almost every Landsman among its supporters from the Kings down.

Every now and then a priest would forget to wear his Skin out-of-doors and be arrested, then released later in an official jail-break. Those who refused to cooperate were forcibly kidnapped, taken to another town and there let loose. Nor did it do the priest any good to proclaim boldly who he was. Everybody pretended not to know he was a fugitive from justice. They insisted on calling him by his official pseudonym.

However, few priests were such martyrs. Generations of Skin-wearing had sapped the ecclesiastical vigor.

The thing that puzzled Rastignac about Father Jules was the sacrament wine. Neither he nor anybody else in L'Bawpfey, as far as he knew, had ever tasted the liquid outside of the ceremony. Indeed, except for certain of the priests, nobody even knew how to make wine.

He shook the priest awake, said, "What's the matter, Father?"

Father Jules burst into tears. "Ah, my boy, you have caught me in my sin. I am a drunkard."

Everybody looked blank. "What does that word *drunkard* mean?"

"It means a man who's damned enough to fill his Skin with alcohol, my boy, fill it until he's no longer a man but a beast."

"Alcohol? What is that?"

"The stuff that's in the wine, my boy. You don't know what I'm talking about because the knowledge was long ago forbidden except to us of the cloth. Cloth, he says! Bah! We go around like everybody, naked except for these extradermal monstrosities which reveal rather than conceal, which not only serve us as clothing but as mentors, parents, censors, interpreters, and, yes, even as priests. Where's a bottle that's not empty? I'm thirsty."

Rastignac stuck to the subject "Why was the making of this alcohol forbidden?"

"How should I know?" said Father Jules. "I'm old, but not so ancient that I came with the Six Flying Stars.... Where is that bottle?"

Rastignac was not offended by his crossness. Priests were notorious for being the most ill-tempered, obstreperous, and unstable of men. They were not at all like the clerics of Earth, whom everybody knew from legend had been sweet-tempered, meek, humble, and obedient to authority. But on L'Bawpfey these men of the Church had reason to be out of sorts. Everybody attended Mass, paid their tithes, went to confession, and did not fall asleep during sermons. Everybody believed what the priests told them and were as good as it was possible for human beings to be. So, the priests had no real incentive to work, no evil to fight.

Then why the prohibition against alcohol?

"*Sacre Bleu!*" groaned Father Jules. "Drink as much as I did last night and you'll find out. Never again, I say. Ah, there's another bottle, hidden by a providential fate under my traveling robe. Where's that corkscrew?"

Father Jules swallowed half of the bottle, smacked his lips, picked up his Skin from the floor, brushed off the dirt and said, "I must be going, my sons. I've a noon appointment with the bishop, and I've a good twelve kilometers to travel. Perhaps one of you gentlemen has a car?"

Rastignac shook his head and said he was sorry but their car was tired and had, besides, thrown a shoe. Father Jules shrugged philosophically, put on his Skin and reached out again for the bottle.

Rastignac said, "Sorry, Father. I'm keeping this bottle."

"For what?" asked father Jules.

"Never mind. Say I'm keeping you from temptation."

"Bless you, my son, and may you have a big enough hangover to show you the wickedness of your ways."

Smiling, Rastignac watched the Father walk out. He was not disappointed. The priest had no sooner reached the huge door than his Skin fell off and lay motionless upon the stone.

"Ah," breathed Rastignac. "The same thing happened to Mapfabvisheen when he put his on. There must be something about the wine that deadens the Skins, makes them fall off."

After the padre had left, Rastignac handed the bottle to Mapfarity. "We're dedicated to breaking the law most illegally, brother. So I'm asking you to analyze this wine and find out how to make it."

"Why not ask Father Jules?"

"Because priests are pledged never to reveal the secret. That was one of the original agreements whereby the Church was allowed to remain on L'Bawpfey. Or, at least that's what my parish priest told me. He said it was a good thing, as it removed an evil from man's temptation. He never did say why it was so evil. Maybe he didn't know.

"That doesn't matter. What does matter is that the Church has inadvertently given us a weapon whereby we may free Man from his bondage to the Skins and it has also given itself once again a chance to be really persecuted and to flourish on the blood of its martyrs."

"Blood?" said Lusine, licking her lips. "The Churchmen drink blood?"

Rastignac did not explain. He could be wrong. If so, he'd feel less like a fool if they didn't know what he thought.

Meanwhile, there were the first steps to be taken for the unskinning of an entire planet.

Chapter IX

Later that day the mucketeers surrounded the castle but they made no effort to storm it. The following day one of them knocked on the huge front door and presented Mapfarity with a summons requiring them to surrender. The Giant laughed, put the document in his mouth and ate it. The server fainted and had to be revived with a bucket of cold water before he could stagger back to report this tradition-shattering reception.

Rastignac set up his underground so it could be expanded in a hurry. He didn't worry about the blockade because, as was well known, Giants' castles had all sorts of subterranean tunnels and secret exits. He contacted a small number of priests who were willing to work for him. These were congenital rebels who became quite enthusiastic when he told them their activities would result in a fierce persecution of the Church.

The majority, however, clung to their Skins and said they would have nothing to do with this extradermal-less devil. They took pride and comfort in that term. The vulgar phrase for the man who refused to wear his Skin was "devil," and, by law and logic, the Church could not be associated with a devil. As everybody knew, the priests have always been on the side of the angels.

Meanwhile, the Devil's band slipped out of the tunnels and made raids. Their targets were Giants' castles and government treasuries; their loot, the geese. So many raids did they make that the president of the League of Giants and the Business Agent for the Guild of Egg-stealers came to plead with them. And remained to denounce. Rastignac was delighted with their complaints, and, after listening for a while, threw them out.

Rastignac had, like all other Skin-wearers, always accepted the monetary system as a thing of reason and steady balance. But, without his Skin he was able to think objectively and saw its weaknesses.

For some cause buried far in history, the Giants had always had control of the means for making the hexagonal golden coins called *oeufs*. But the Kings, wishing to get control of the golden eggs, had set up that élite branch of the Guild which specialized in abducting the half-living 'geese.' Whenever a thief was successful he turned the goose over to his King. The monarch, in turn, sent a note to the robbed Giant informing him that the government intended to keep the goose to make its own currency. But even though the Giant was making counterfeit geese, the King, in his generosity, would ship to the Giant one out of every thirty eggs laid by the kidnappee.

The note was a polite and well-recognized lie. The Giants made the only genuine gold-egg-laying geese on the planet because the Giants' League alone knew the secret. And the King gave back one-thirtieth of his loot so the Giant could accumulate enough money to buy the materials to create another goose. Which would, possibly, be stolen later on.

Rastignac, by his illegal rape of geese, was making money scarce. Peasants were hanging on to their produce and waiting to sell until prices were at their highest. The government, merchants, the league, the guild, all saw themselves impoverished.

Furthermore, the Amphibs, taking note of the situation, were making raids of their own and blaming them on Rastignac.

He did not care. He was intent on trying to find a way to reach Kataproimnoin and rescue the Earthman so he could take off in the spaceship floating in the harbor. But he knew that he would have to take things slowly, to scout out the land and plan accordingly.

Furthermore, Mapfarity had made him promise he would do his best to set up the Landsmen so they would be able to resist the Waterfolk when the day for war came.

Rastignac made his biggest raid when he and his band stole one moonless night into the capital itself to rob the big Goose House, only an egg's throw away from the Palace and the Ministry of Ill-Will. They put the Goose House guards to sleep with little arrows smeared with dream-snake venom, filled their lead-leaf-lined bags with gold eggs, and sneaked out the back door.

As they left, Rastignac saw a cloaked figure slinking from the back door of the Ministry. Seized with intuition, he tackled the figure. It was an Amphib-changeling. Rastignac struck the Amphib with a venomous arrow before the Water-human could cry out or stab back.

Mapfarity grabbed up the limp Amphib and they raced for the safety of the castle.

They questioned the Amphib, Pierre Pusipremnoos, in the castle. At first silent, he later began talking freely when Mapfarity got a heavy Skin from his fleshforge and put it on the fellow. It was a Skin modeled after those worn by the Water-people, but it differed in that the Giant could control, through another Skin, the powerful neural shocks.

After a few shocks Pierre admitted he was the foster-son of the Amphibian King and that, incidentally, Lusine was his foster-sister. He further stated he was a messenger between the Amphib King and the Ssarraror's Ill-Will Minister.

More shocks extracted the fact that the Minister of Ill-Will, Auverpin, was an Amphib-changeling who was passing himself off as a born Landsman. Not only that, the Human hostages among the Amphibs were about to stage a carefully planned revolt against the born Amphibs. It would kill off about half of them. The rest would then be brought under control of the Master Skin.

When the two stepped from the lab they were attacked by Lusine, knife in hand. She gashed Rastignac in the arm before he knocked her out with an upper-cut. Later, while Mapfarity applied a little jelly-like creature called a *scar-jester* to the wound, Rastignac complained:

"I don't know if I can endure much more of this. I thought the way of Violence would not be hard to follow because I hated the Skins and the Amphibs so much. But it is easier to attack a faceless, hypothetical enemy, or torture him, than the individual enemy. Much easier."

"My brother," boomed the Giant, "if you continue to dwell upon the philosophical implications of your actions you will end up as helpless and confused as the leg-counting centipede. Better not think. Warriors are not supposed to. They lose their keen fighting edge when they think. And you need all of that now."

"I would suppose that thought would sharpen them."

"When issues are simple, yes. But you must remember that the system on this planet is anything but uncomplicated. It was set up to confuse, to keep one always off balance. Just try to keep one thing in mind—the Skins are far more of an impediment to Man than they are a help. Also, that if the Skins don't come off the Amphibs will soon be cutting our throats. The only way to save ourselves is to kill them first. Right?"

"I suppose so," said Rastignac. He stooped and put his hands under the unconscious Lusine's armpits. "Help me put her in a room. We'll keep her locked up until she cools off. Then we'll use her to guide us when we get to Kataproimnoin. Which reminds me—how many gallons of the wine have you made so far?"

Chapter X

A week later Rastignac summoned Lusine. She came in frowning, and with her lower lip protruding in a pretty pout.

He said, "Day after tomorrow is the day on which the new Kings are crowned, isn't it?"

Tonelessly she said, "Supposedly. Actually, the present Kings will be crowned again."

Rastignac smiled. "I know. Peculiar, isn't it, how the 'people' always vote the same Kings back into power? However, that isn't what I'm getting at. If I remember correctly, the Amphibs give their King exotic and amusing gifts on coronation day. What do you think would happen if I took a big shipload of bottles of wine and passed it out among the population just before the Amphibs begin their surprise massacre?"

Lusine had seen Mapfarity and Rastignac experimenting with the wine and she had been frightened by the results. Nevertheless, she made a brave attempt to hide her fear now. She spit at him and said, "You mud-footed fool! There are priests who will know what it is! They will be in the coronation crowd."

"Ah, not so! In the first place, you Amphibs are almost entirely Aggressive Pantheists. You have only a few priests, and you will now pay for that omission of wine-tasters. Second, Mapfarity's concoction tastes not at all vinous and is twice as strong."

She spat at him again and spun on her heel and walked out.

That night Rastignac's band and Lusine went through a tunnel which brought them up through a hollow tree about two miles west of the castle. There they hopped into the Renault, which had been kept in a camouflaged garage, and drove to the little port of Marrec. Archambaud had paved their way here with golden eggs and a sloop was waiting for them.

Rastignac took the boat's wheel. Lusine stood beside him, ready to answer the challenge of any Amphib patrol that tried to stop them. As the Amphib-King's foster-daughter, she could get the boat through to the Amphib island without any trouble at all.

Archambaud stood behind her, a knife under his cloak, to make sure she did not try to betray them. Lusine had sworn she could be trusted. Rastignac had answered that he was sure she could be, too, as long as the knife point pricked her back to remind her.

Nobody stopped them. An hour before dawn they anchored in the harbor of Kataproimnoin. Lusine was tied hand and foot inside the cabin. Before Rastignac could scratch her with dream-snake venom, she pleaded, "You could not do this to me, Jean-Jacques, if you loved me."

"Who said anything about loving you?"

"Well, I like that! You said so, you cheat!"

"Oh, *then*! Well, Lusine, you've had enough experience to know that such protestations of tenderness and affection are only inevitable accompaniments of the moment's passion."

For the first time since he had known her he saw Lusine's lower lip tremble and tears come in her eyes. "Do you mean you were only using me?" she sobbed.

"You forget I had good reason to think you were just using *me*. Remember, you're an Amphib, Lusine. Your people can't be trusted. You blood-drinkers are as savage as the little sea-monsters you leave in Human cradles."

"Jean-Jacques, take me with you! I'll do anything you say! I'll even cut my foster-father's throat for you!"

He laughed. Unheeding, she swept on. "I want to be with you, Jean-Jacques! Look, with me to guide you in, my homeland—with my prestige as the Amphib-King's daughter—you can become King yourself after the rebellion. I'd get rid of the Amphib-King for you so there'll be nobody in your way!"

She felt no more guilt than a tigress. She was naive and terrible, innocent and disgusting.

"No, thanks, Lusine." He scratched her with the dream-snake needle. As her eyes closed he said, "You don't understand. All I want to do is voyage to the stars. Being King means nothing to me. The only person I'd trade places with would be the Earthman the Amphibs hold prisoner."

He left her sleeping in the locked cabin.

Noon found them loafing on the great square in front of the Palace of the Two Kings of the Sea and the Islands. All were disguised as Waterfolk. Before they'd left the castle, they had grafted webs between their fingers and toes—just as Amphib-changelings who weren't born with them, did—and they wore the special Amphib Skins that Mapfarity had grown in his fleshforge. These were able to tune in on the Amphibs' wavelengths, but they lacked their shock mechanism.

Rastignac had to locate the Earthman, rescue him, and get him to the spaceship that lay anchored between two wharfs, its sharp nose pointing outwards. A wooden bridge had been built from one of the wharfs to a place halfway up its towering side.

Rastignac could not make out any breaks in the smooth metal that would indicate a port, but reason told him there must be some sort of entrance to the ship at that point.

A guard of twenty Amphibs repulsed any attempt on the crowd's part to get on the bridge.

Rastignac had contacted the harbor-master and made arrangements for workmen to unload his cargo of wine. His freehandedness with the gold eggs got him immediate service even on this general holiday. Once in the square, he and his men uncrated the wine but left the two heavy chests on the wagon which was hitched to a powerful little six-legged Jeep.

They stacked the bottles of wine in a huge pile while the curious crowd in the square encircled them to watch. Rastignac then stood on a chest to survey the scene, so that he could best judge the time to start. There were perhaps seven or eight thousand of all three races there—the Ssassarors, the Amphibs, the Humans—with an unequal portioning of each.

Rastignac, looking for just such a thing, noticed that every non-human Amphib had at least two Humans tagging at his heels.

It would take two Humans to handle an Amphib or a Ssassaror. The Amphibs stood upon their seal-like hind flippers at least six and a half feet tall and weighed about three hundred pounds. The Giant Ssassarors, being fisheaters, had reached the same enormous height as Mapfarity. The Giants were in the minority, as the Amphibs had always preferred stealing Human babies from the Terrans. These were marked for death as much as the Amphibs.

Rastignac watched for signs of uneasiness or hostility between the three groups. Soon he saw the signs. They were not plentiful, but they were enough to indicate an uneasy undercurrent. Three times the guards had to intervene to break up quarrels. The Humans eyed the non-human quarrelers, but made no move to help their Amphib fellows against the Giants. Not only that, they took them aside afterwards and seemed to be reprimanding them. Evidently the order was that everyone was to be on his behavior until the time to revolt. Rastignac glanced at the great tower-clock. "It's an hour before the ceremonies begin," he said to his men. "Let's go."

Chapter XI

Mapfarity, who had been loitering in the crowd some distance away, caught Archambaud's signal and slowly, as befit a Giant whose feet hurt, limped towards them. He stopped, scrutinized the pile of bottles, then, in his lion's-roar-at-the-bottom-of-a-well voice said, "Say, what's in these bottles?"

Rastignac shouted back, "A drink which the new Kings will enjoy very much."

"What's that?" replied Mapfarity. "Sea-water?"

The crowd laughed.

"No, it's not water," Rastignac said, "as anybody but a lumbering Giant should know. It is a delicious drink that brings a rare ecstacy upon the drinker. I got the formula for it from an old witch who lives on the shores of far off Apfelabvidanahyew. He told me it had been in his family since the coming of Man to L'Bawpfey. He parted with the formula on condition I make it only for the Kings."

"Will only Their Majesties get to taste this exquisite drink?" bellowed Mapfarity.

"That depends upon whether it pleases Their Majesties to give some to their subjects to celebrate the result of the elections."

Archambaud, also planted in the crowd, shrilled, "I suppose if they do, the big-paunched Amphibs and Giants will get twice as much as us Humans. They always do, it seems."

There was a mutter from the crowd; approbation from the Amphibs, protest from the others.

"That will make no difference," said Rastignac, smiling. "The fascinating thing about this is that an Amphib can drink no more than a Human. That may be why the old man who revealed his secret to me called the drink Old Equalizer."

"Ah, you're skinless," scoffed Mapfarity, throwing the most deadly insult known. "I can out-drink, out-eat, and out-swim any Human here. Here, Amphib, give me a bottle, and we'll see if I'm bragging."

An Amphib captain pushed himself through the throng, waddling clumsily on his flippers like an upright seal.

"No, you don't!" he barked. "Those bottles are intended for the Kings. No commoner touches them, least of all a Human and a Giant."

Rastignac mentally hugged himself. He couldn't have planned a better intervention himself! "Why can't I?" he replied. "Until I make an official presentation, these bottles are mine, not the Kings'. I'll do what I want with them."

"Yeah," said the Amphibs. "That's telling him!"

The Amphib's big brown eyes narrowed and his animal-like face wrinkled, but he couldn't think of a retort. Rastignac at once handed a bottle apiece to each of his comrades. They uncorked and drank and then assumed an ecstatic expression which was a tribute to their acting, for these three bottles held only fruit juice.

"Look here, captain," said Rastignac, "why don't you try a swig yourself? Go ahead. There's plenty. And I'm sure Their Majesties would be pleased to contribute some of it on this joyous occasion. Besides, I can always make more for the Kings.

"As a matter of fact," he added, winking, "I expect to get a pension from the courts as the Kings' Old Equalizer-maker."

The crowd laughed. The Amphib, afraid of losing face, took the bottle—which contained wine rather than fruit juice. After a few long swallows the Amphib's eyes became red and a silly grin curved his thin, black-edged lips. Finally, in a thickening voice, he asked for another bottle.

Rastignac, in a sudden burst of generosity, not only gave him one, but began passing out bottles to the many eager reaching hands. Mapfarity and the two egg-thieves helped him. In a short time, the pile of bottles had dwindled to a fourth of its former height. When a mixed group of guards strode up and demanded to know what the commotion was about, Rastignac gave them some of the bottles.

Meanwhile, Archambaud slipped off into the mob. He lurched into an Amphib, said something nasty about his ancestors, and pulled his knife. When the Amphib lunged for the little man, Archambaud jumped back and shoved a Human-Amphib into the giant flipper-like arms.

Within a minute the square had erupted into a fighting mob. Staggering, red-eyed, slur-tongued, their long-repressed hostility against each other, released by the liquor which their bodies were unaccustomed to, Human, Ssassaror and Amphib fell to with the utmost will, slashing, slugging, fighting with everything they had.

None of them noticed that every one who had drunk from the bottles had lost his Skin. The Skins had fallen off one by one and lay motionless on the pavement where they were kicked or stepped upon. Not one Skin tried to crawl back to its owner because they were all nerve-numbed by the wine.

Rastignac, seated behind the wheel of the Jeep, began driving as best he could through the battling mob. After frequent stops he halted before the broad marble steps that ran like a stairway to heaven, up and up before it ended on the Porpoise Porch of the Palace. He and his gang were about to take the two heavy chests off the wagon when they were transfixed by a scene before them.

A score of dead Humans and Amphibs lay on the steps, evidence of the fierce struggle that had taken place between the guards of the two monarchs. Evidently the King had heard of the riot and hastened outside. There the Amphib-changeling King had apparently realized that the rebellion was way ahead of schedule, but he had attacked the Amphib King anyway.

And he had won, for his guardsmen held the struggling flipper-footed Amphib ruler down while two others bent his head back over a step. The Changeling-King himself, still clad in the coronation robes, was about to draw his long ceremonial knife across the exposed and palpitating throat of the Amphib King.

This in itself was enough to freeze the onlookers. But the sight of Lusine running up the stairway towards the rulers added to their paralysis. She had a knife in her hand and was holding it high as she ran toward her foster-father, the Amphib King.

Mapfarity groaned, but Rastignac said, "It doesn't matter that she has escaped. We'll go ahead with our original plan."

They began unloading the chests while Rastignac kept an eye on Lusine. He saw her run up, stop, say a few words to the Amphib King, then kneel and stab him, burying the knife in his jugular vein. Then, before anybody could stop her she had applied her mouth to the cut in his neck.

The Human-King kicked her in the ribs and sent her rolling down the steps. Rastignac saw correctly that it was not her murderous deed that caused his reaction. It was because she had dared to commit it without his permission and had also drunk the royal blood first.

He further noted with grim satisfaction that when Lusine recovered from the blow and ran back up to talk to the King, he ignored her. She pointed at the group around the wagon but he dismissed her with a wave of his hand. He was too busy gloating over his vanquished rival lying at his feet.

The plotters hoisted the two chests and staggered up the steps. The King passed them as he went down with no more than a curious glance. Gifts had been coming up those steps all day for the King, so he undoubtedly thought of them only as more gifts. So Rastignac and his men walked past the knives of the guards as if they had nothing to fear.

Lusine stood alone at the top of the steps. She was in a half-crouch, knife ready. "I'll kill the King and I'll drink from his throat!" she cried hoarsely. "No man kicks me except for love. Has he forgotten that I am the foster-daughter of the Amphib King?"

Rastignac felt revulsion but he had learned by now that those who deal in violence and rebellion must march with strange steppers.

"Bear a hand here," he said, ignoring her threat.

Meekly she grabbed hold of a chest's corner. To his further questioning, she replied that the Earthman who had landed in the ship was held in a suite of rooms in the west wing. Their trip thereafter was fast and direct. Unopposed, they carted the chests to the huge room where the Master Skin was kept.

There they found ten frantic bio-technicians excitedly trying to determine why the great extraderm—the Master Skin through which all individual Skins were controlled—was not broadcasting properly. They had no way as yet of knowing that it was operating perfectly but that the little Skins upon the Amphibs and their hostage Humans were not shocking them into submission because they were lying in a wine-stupor on the ground. No one had told them that the Skins, which fed off the bloodstream of their hosts, had become anesthetized from the alcohol and failed any longer to react to their Master Skin.

That, of course, applied only to those Skins in the square that were drunk from the wine. Elsewhere all over the kingdom, Amphibs writhed in agony and Ssassarors and Terrans were taking advantage of their helplessness to cut their throats. But not here, where the crux of the matter was.

Chapter XII

The Landsmen rushed the techs and pushed them into the great chemical vat in which the twenty-five hundred foot square Master Skin floated. Then they uncrated the lead-leaf-lined bags filled with stolen geese and emptied them into the nutrient fluid. According to Mapfarity's calculations, the radio-activity from the silicon-carbon geese should kill the big Skin within a few days. When a new one was grown, that, too, would die. Unless the Amphib guessed what was wrong and located the geese on the bottom of the ten-foot deep tank, they would not be able to stop the process. That did not seem likely.

In either case, it was necessary that the Master Skin be put out of temporary commission, at least, so the Amphibs over the Kingdom could have a fighting chance. Mapfarity plunged a hollow harpoon into the isle of floating protoplasm and through a tube connected to that poured into the Skin three gallons of the dream-snake venom. That was enough to knock it out for an hour or two. Meanwhile, if the Amphibs had any sense at all, they'd have rid themselves of their extraderms.

They left the lab and entered the west wing. As they trotted up the long winding corridors Lusine said, "Jean-Jacques, what do you plan on doing now? Will you try to make yourself King of the Terrans and fight us Amphibs?" When he said nothing she went on. "Why don't you kill the Amphib-changeling King and take over here? I could help you do that. You could then have all of L'Bawpfey in your power."

He shot her a look of contempt and cried, "Lusine, can't you get it through that thick little head of yours that everything I've done has been done so that I can win one goal: reach the Flying Stars? If I can get the Earthman to his ship I'll leave with him and not set foot again for years on this planet. Maybe never again."

She looked stricken. "But what about the war here?" she asked.

"There are a few men among the Landfolk who are capable of leading in wartime. It will take strong men, and there are very few like me, I admit, but—oh, oh, opposition!" He broke off at sight of the six guards who stood before the Earthman's suite.

Lusine helped, and within a minute they had slain three and chased away the others. Then they burst through the door—and Rastignac received another shock.

The occupant of the apartment was a tiny and exquisitely formed redhead with large blue eyes and very unmasculine curves!

"I thought you said Earth*man*?" protested Rastignac to the Giant who came lumbering along behind them.

"Oh, I used that in the generic sense," Mapfarity replied. "You didn't expect me to pay any attention to sex, did you? I'm not interested in the gender of you Humans, you know."

There was no time for reproach. Rastignac tried to explain to the Earthwoman who he was, but she did not understand him. However, she did seem to catch on to what he wanted and seemed reassured by his gestures. She picked up a large book from a table and, hugging it to her small, high and rounded bosom, went with him out the door.

They raced from the palace and descended onto the square. Here they found the surviving Amphibs clustered into a solid phalanx and fighting, bloody step by step, towards the street that led to the harbor.

Rastignac's little group skirted the battle and started down the steep avenue toward the harbor. Halfway down he glanced back and saw that nobody as yet was paying any attention to them. Nor was there anybody on the street to bother them, though the pavement was strewn with Skins and bodies. Apparently, those who'd lived through the first savage mêlée had gone to the square.

They ran onto the wharf. The Earthwoman motioned to Rastignac that she knew how to open the spaceship, but the Amphibs didn't. Moreover, if they did get in, they wouldn't know how to operate it. She had the directions for so doing in the book hugged so desperately to her chest. Rastignac surmised she hadn't told the Amphibs about that. Apparently they hadn't, as yet, tried to torture the information from her.

Therefore, her telling him about the book indicated she trusted him.

Lusine said, "Now what, Jean-Jacques? Are you still going to abandon this planet?"

"Of course," he snapped.

"Will you take me with you?"

He had spent most of his life under the tutelage of his Skin, which ensured that others would know when he was lying. It did not come easy to hide his true feelings. So a habit of a lifetime won out.

"I will not take you," he said. "In the first place, though you may have some admirable virtues, I've failed to detect one. In the second place, I could not stand your blood-drinking nor your murderous and totally immoral ways."

"But, Jean-Jacques, I will give them up for you!"

"Can the shark stop eating fish?"

"You would leave Lusine, who loves you as no Earthwoman could, and go with that—that pale little doll I could break with my hands?"

"Be quiet," he said. "I have dreamed of this moment all my life. Nothing can stop me now."

They were on the wharf beside the bridge that ran up the smooth side of the starship. The guard was no longer there, though bodies showed that there had been reluctance on the part of some to leave.

They let the Earthwoman precede them up the bridge.

Lusine suddenly ran ahead of him, crying, "If you won't have me, you won't have her, either! Nor the stars!"

Her knife sank twice into the Earthwoman's back. Then, before anybody could reach her, she had leaped off the bridge and into the harbor.

Rastignac knelt beside the Earthwoman. She held out the book to him, then she died. He caught the volume before it struck the wharf.

"My God! My God!" moaned Rastignac, stunned with grief and shock and sorrow. Sorrow for the woman and shock at the loss of the ship and the end of his plans for freedom.

Mapfarity ran up then and took the book from his nerveless hand. "She indicated that this is a manual for running the ship," he said. "All is not lost."

"It will be in a language we don't know," Rastignac whispered.

Archambaud came running up, shrilled, "The Amphibs have broken through and are coming down the street! Let's get to our boat before the whole blood-thirsty mob gets here!"

Mapfarity paid him no attention. He thumbed through the book, then reached down and lifted Rastignac from his crouching position by the corpse.

"There's hope yet, Jean-Jacques," he growled. "This book is printed with the same characters as those I saw in a book owned by a priest I knew. He said it was in Hebrew, and that it was the Holy Book in the original Earth language. This woman must be a citizen of the Republic of Israeli, which I understand was rising to be a great power on Earth at the time you French left.

"Perhaps the language of this woman has changed somewhat from the original tongue, but I don't think the alphabet has. I'll bet that if we get this to a priest who can read it—there are only a few left—he can translate it well enough for us to figure out everything."

They walked to the wharf's end and climbed down a ladder to a platform where a dory was tied up. As they rowed out to their sloop Mapfarity said:

"Look, Rastignac, things aren't as bad as they seem. If you haven't the ship nobody else has, either. And you alone have the key to its entrance and operation. For that you can thank the Church, which has preserved the ancient wisdom for emergencies which it couldn't forsee, such as this. Just as it kept the secret of wine, which will eventually be the greatest means for delivering our people from their bondage to the Skins and, thus enable them to fight the Amphibs back instead of being slaughtered.

"Meanwhile, we've a battle to wage. You will have to lead it. Nobody else but the Skinless Devil has the prestige to make the people gather around him. Once we accuse the Minister of Ill-Will of treason and jail him, without an official Breaker to release him, we'll demand a general election. You'll be made King of the Ssassaror; I, of the Terrans. That is inevitable, for we are the only skinless men and, therefore, irresistible. After the war is won, we'll leave for the stars. How do you like that?"

Rastignac smiled. It was weak, but it was a smile. His bracket-shaped eyebrows bent into their old sign of determination.

"You are right," he replied. "I have given it much thought. A man has no right to leave his native land until he's settled his problems here. Even if Lusine hadn't killed the Earthwoman and I had sailed away, my conscience wouldn't have given me any rest. I would have known I had abandoned the fight in the middle of it. But now that I have stripped myself of my Skin—which was a substitute for a conscience—and now that I am being forced to develop my own inward conscience, I must admit that immediate flight to the stars would have been the wrong thing."

The pleased and happy Mapfarity said, "And you must also admit, Rastignac, that things so far have had a way of working out for the best.

Even Lusine, evil as she was, has helped towards the general good by keeping you on this planet. And the Church, though it has released once again the old evil of alcohol, has done more good by so doing than...."

But here Rastignac interrupted to say he did not believe in this particular school of thought, and so, while the howls of savage warriors drifted from the wharfs, while the structure of their world crashed around them, they plunged into that most violent and circular of all whirlpools—the Discussion Philosophical.